Sorting Out

Worms

and Other Invertebrates

Everything You Want to Know About Insects, Corals, Mollusks, Sponges, and More!

Samuel G. Woods

BLACKBIRCH PRESS, INC.

WOODBRIDGE, CONNECTICUT

Published by Blackbirch Press, Inc.
260 Amity Road
Woodbridge, CT 06525

©1999 by Blackbirch Press, Inc.
First Edition

e-mail: staff@blackbirch.com
Web site: www.blackbirch.com

Printed in Hong Kong

10 9 8 7 6 5 4 3 2 1

Photo Credits
Cover: all images ©Corel Corporation, except worm ©Kerry L. Werry; pages 4-16, 17 (top and bottom), 18-25, 28: ©Corel Corporation; pages 17 (middle), 26, 27: ©PhotoDisc; pages 27 (bottom), 29, 30: ©Kerry L. Werry.

Library of Congress Cataloging-in-Publication Data
Woods, Samuel G.
Sorting out worms and other invertebrates : everything you want to know about insects, corals, mollusks, sponges, and more! / Samuel G. Woods
  p. cm. — (Sorting out)
  Includes bibliographical references.
  Summary: Text and photographs introduce such invertebrates as insects, spiders, crustaceans, jellyfish, sea anemones, sand dollars, squid, sponges, flatworms, and more.
  ISBN 1-56711-371-0
  1. Invertebrates—Juvenile literature. [1. Invertebrates.] I. Title. II. Series.
QL362.4.W66 1999
592—dc21
                         99-31237
                            CIP
                            AC

Contents

Note: *For an explanation of how living things (especially invertebrates) are classified (organized), turn to the chart on page 31.*

Insects (Arthropods)

Scientific Name: *Phylum Arthropoda*
How to Say It: AR•THRO•PO•DAH
Smaller Group (Class) Within Arthropoda: Insecta

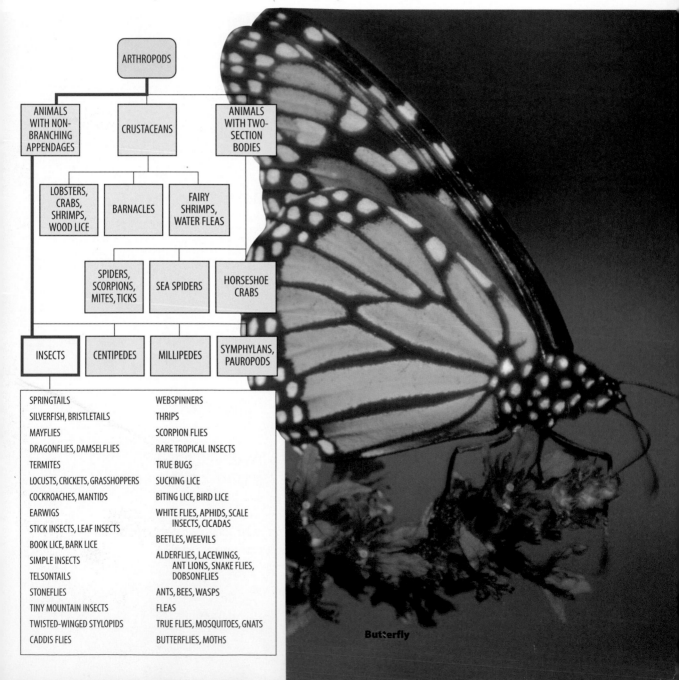

ARTHROPODS

ANIMALS WITH NON-BRANCHING APPENDAGES

CRUSTACEANS

ANIMALS WITH TWO-SECTION BODIES

LOBSTERS, CRABS, SHRIMPS, WOOD LICE

BARNACLES

FAIRY SHRIMPS, WATER FLEAS

SPIDERS, SCORPIONS, MITES, TICKS

SEA SPIDERS

HORSESHOE CRABS

INSECTS

CENTIPEDES

MILLIPEDES

SYMPHYLANS, PAUROPODS

SPRINGTAILS

SILVERFISH, BRISTLETAILS

MAYFLIES

DRAGONFLIES, DAMSELFLIES

TERMITES

LOCUSTS, CRICKETS, GRASSHOPPERS

COCKROACHES, MANTIDS

EARWIGS

STICK INSECTS, LEAF INSECTS

BOOK LICE, BARK LICE

SIMPLE INSECTS

TELSONTAILS

STONEFLIES

TINY MOUNTAIN INSECTS

TWISTED-WINGED STYLOPIDS

CADDIS FLIES

WEBSPINNERS

THRIPS

SCORPION FLIES

RARE TROPICAL INSECTS

TRUE BUGS

SUCKING LICE

BITING LICE, BIRD LICE

WHITE FLIES, APHIDS, SCALE INSECTS, CICADAS

BEETLES, WEEVILS

ALDERFLIES, LACEWINGS, ANT LIONS, SNAKE FLIES, DOBSONFLIES

ANTS, BEES, WASPS

FLEAS

TRUE FLIES, MOSQUITOES, GNATS

BUTTERFLIES, MOTHS

Butterfly

Rhino beetle

Earwig

Bee

Weevil

NOTEPAD

Instead of bones inside their bodies, arthropods have hard shells on their outsides called exoskeletons. In order to grow larger, arthropods need to molt (shed) their exoskeletons and grow new ones.

Insects are just one kind of arthropod. Insects have bodies divided into three sections: head, thorax (midsection), and abdomen (rear section). They also have six jointed legs.

Noteworthy: Arthropods make up about 75% of all known animal species!

5

Sea Stars, Feather Stars, and Others (Echinoderms)

Scientific Name: *Echinoderm (Phylum Echinodermata)*
How to Say It: EE•KYNE•O•DURM

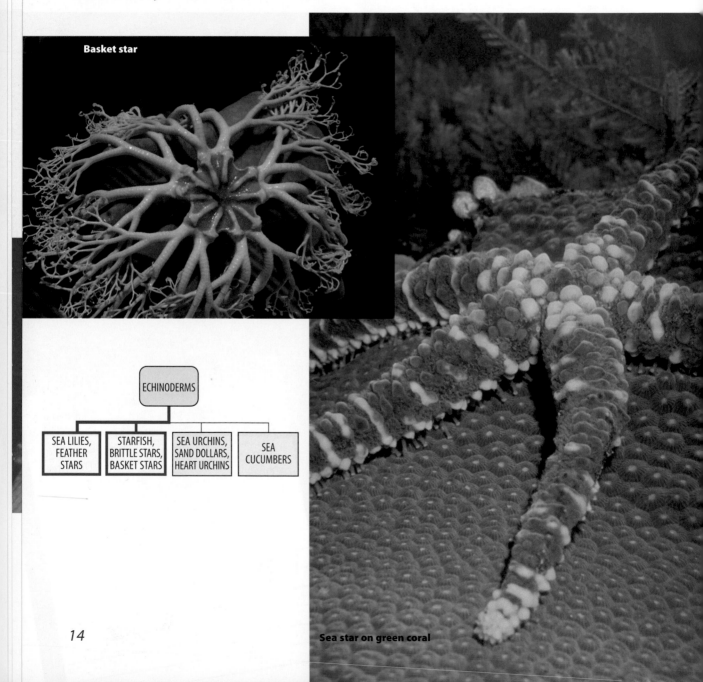

Basket star

ECHINODERMS

- SEA LILIES, FEATHER STARS
- STARFISH, BRITTLE STARS, BASKET STARS
- SEA URCHINS, SAND DOLLARS, HEART URCHINS
- SEA CUCUMBERS

14

Sea star on green coral

Feather star

Brittle star on orange sponge

NOTEPAD

These invertebrates are broken down into four smaller groupings, or classes. The four classes include: sea lilies and feather stars; sea stars, brittle stars, and basket stars; sea urchins, sand dollars, and heart urchins; sea cucumbers. Echinoderms live on the sea floor and have bodies that are shaped in five-point patterns. Their external skeletons lie just beneath an outer layer of skin. They have tube feet and no heads.